RAILWAY INDUSTRY STANDARD
OF THE PEOPLE'S REPUBLIC OF CHINA

Technical Specification for Heat Treated Rails

TB/T 2635-2004

Issued by Ministry of Railways of the People's Republic of China

Issued Date: January 30, 2004

Vaild Date: March 1, 2004

China Railway Publishing House

Beijing 2015

图书在版编目(CIP)数据

热处理钢轨技术条件：TB/T 2635-2004：英文/中华人民共和国国家铁路局组织编译. —北京：中国铁道出版社，2015.4
ISBN 978-7-113-20027-5

Ⅰ.①热… Ⅱ.①中… Ⅲ.①钢轨—热处理—技术条件—英文 Ⅳ.①TG162.82

中国版本图书馆CIP数据核字(2015)第040036号

Chinese version first published in the People's Republic of China in 2004
English version first published in the People's Republic of China in 2015
by China Railway Publishing House
No. 8, You'anmen West Street, Xicheng District
Beijing, 100054
www.tdpress.com

Printed in China by Printing Factory of China Railway Publishing House

© 2004 by Ministry of Railways of the People's Republic of China

All rights reserved. No part of this publication may be reproduced or transmitted in any form or by any means, electronic or mechanical, including photocopying, recording, or by any information storage and retrieval systems, without the prior written consent of the publisher.

This book is sold subject to the condition that it shall not, by way of trade or otherwise, be lent, resold, hired out or otherwise circulated without the publisher's prior consent in any form of binding or cover other than that in which it is published and without a similar condition including this condition being imposed on the subsequent purchaser.

ISBN 978-7-113-20027-5

Introduction to the English Version

To promote the exchange and cooperation in railway technology between China and the rest of the world, China Academy of Railway Sciences, entrusted by National Railway Administration of the People's Republic of China and organized by CHINA RAILWAY, conducted the translation and preparation of Chinese railway technology and product standards.

This standard is the official English language version of *Technical Specification for Heat Treated Rails* (TB/T 2635-2004). The original Chinese version of this standard was issued by the former Ministry of Railways of the People's Republic of China and came into effect on March 1, 2004. In case of discrepancies between the two versions, the Chinese version shall prevail. National Railway Administration of the People's Republic of China owns the copyright of this English version.

The English version was prepared by China Academy of Railway Sciences.

Your comments are invited for next revision of this standard and should be addressed to Technology and Legislation Department of National Railway Administration and China Academy of Railway Sciences.

Address: Technology and Legislation Department of National Railway Administration, No. 6, Fuxing Road, Beijing, 100891, P. R. China.

China Academy of Railway Sciences, No. 2, Daliushu Road, Haidian District, Beijing, 100081, P. R. China.

Email: liudalei0912@126.com

The translation was performed by Chen Qi, Han Xu, Liu Dalei, Chen Min, Li Hanqiu, Qian Jun.

The translation was reviewed by Wang Yanhua and Jia Guoping.

Notice on the Issuance of the English Version of Twenty-four Railway Technical Standards including *Rails for High Speed Railway* by National Railway Administration

Document Guo Tie Ke Fa [2015] No. 6

To promote the exchange and cooperation in railway technology between China and the rest of the world, National Railway Administration organized the translation and preparation of twenty-four Chinese railway technical standards including *Rails for High Speed Railway* (TB/T 3276-2011). In case of discrepancies between the original Chinese version and the English version, the Chinese version shall prevail.

The English version is published and distributed by China Railway Publishing House.

Attached here is a list of the English version of these technical standards.

S/N	Chinese title	English title	Standard number
1	高速铁路用钢轨	Rails for High Speed Railway	TB/T 3276-2011
2	热处理钢轨技术条件	Technical Specification for Heat Treated Rails	TB/T 2635-2004
3	钢轨伤损分类	Catalogue of Rail Defects	TB/T 1778-2010
4	电力机车通用技术条件	General Technical Specifications for Electric Locomotives	GB/T 3317-2006
5	电力机车制成后投入使用前的试验方法	Testing of Electric Locomotive on Completion of Construction and before Entry into Service	GB/T 3318-2006
6	机车车辆动力学性能台架试验方法	Tests for Dynamics Performance of Locomotives and Rolling Stocks on Test Rig	TB/T 3115-2005
7	机车车体静强度试验规范	Specifications for Static Strength Test on Car Bodies of Locomotives	TB/T 2541-2010
8	机车转向架技术条件	Technical Specifications of Bogies for Locomotives	GB/T 25332-2010
9	铁道机车车体技术条件 第1部分:内燃机车车体	Technical Specifications for Railway Locomotive Car Body Part 1: Diesel Locomotive Car Body	GB/T 25334.1-2010
10	铁道机车车体技术条件 第2部分:电力机车车体	Technical Specifications for Railway Locomotive Car Body Part 2: Electric Locomotive Car Body	GB/T 25334.2-2010
11	铁道客车通用技术条件	General Technical Specification for Railway Passenger Car	GB/T 12817-2004
12	铁道客车组装后的检查与试验规则	Rules for Inspecting and Testing of Railway Passenger Car after Completion of Construction	GB/T 12818-2004
13	铁道客车通过最小半径曲线试验	Tests on Railway Passenger Cars Negotiating Minimum Radius Curves	TB/T 2218-2010
14	动车组牵引电动机技术条件	Specifications for Traction Motors of Multiple Units	TB/T 3238-2010
15	动车组用内装材料阻燃技术条件	Flame Retardant Technical Specification of Decorating Materials for Multiple Units	TB/T 3237-2010

S/N	Chinese title	English title	Standard number
16	铁路调度通信系统 第1部分:技术条件	Railway Traffic Control Communication System Part 1: Technical Specifications	TB/T 3160.1-2007
17	铁路调度通信系统 第2部分:试验方法	Railway Traffic Control Communication System Part 2: Test Methods	TB/T 3160.2-2007
18	铁路综合接地系统测量方法	Measuring Methods for Railway Integrated Grounding System	TB/T 3233-2010
19	电气化铁道用铜及铜合金接触线	Copper and Copper Alloy Contact Wires for Electrified Railways	TB/T 2809-2005
20	电气化铁道用铜及铜合金绞线	Copper and Copper Alloy Stranded Wires for Electrified Railways	TB/T 3111-2005
21	电气化铁路接触网钢支柱 第1部分:格构式支柱	Steel Poles for Overhead Contact System of Electrified Railways Part 1: Lattice Pole	GB/T 25020.1-2010
22	电气化铁路接触网钢支柱 第2部分:方形钢管支柱	Steel Poles for Overhead Contact System of Electrified Railways Part 2: Square Steel Tube Pole	GB/T 25020.2-2010
23	电气化铁路接触网钢支柱 第3部分:环形钢管支柱	Steel Poles for Overhead Contact System of Electrified Railways Part 3: Ring Steel Tube Pole	GB/T 25020.3-2010
24	电气化铁路接触网钢支柱 第4部分:H形支柱	Steel Poles for Overhead Contact System of Electrified Railways Part 4: H Type Steel Pole	GB/T 25020.4-2010

National Railway Administration of the People's Republic of China

February 16, 2015

Contents

Foreword ·· II
1　Scope ··· 1
2　Normative References ·· 1
3　Terms and Definitions ·· 1
4　Technical Specifications ··· 1
　4.1　Rails ·· 1
　4.2　Chemical Composition ··· 1
　4.3　Profile of Hardened Layer ··· 2
　4.4　Depth of the Hardened Layer ··· 2
　4.5　Hardness ·· 2
　4.6　Microstructure ·· 3
　4.7　Tensile Performance ·· 4
　4.8　External Shape and Allowable Tolerances ·· 4
　4.9　Surface Quality ··· 4
5　Test Method ·· 4
　5.1　Sampling ·· 4
　5.2　Testing Content, Sampling Locations and Quantity of Heat Treated Rail Samples ········· 5
　5.3　Test for Chemical Composition Analysis ·· 5
　5.4　Test for Profile and Depth of the Hardened Layer ·· 5
　5.5　Hardness Test ··· 5
　5.6　Tensile Test ·· 6
　5.7　Microstructure Test ·· 6
　5.8　Inspection for External Shape and Surface Quality ·· 6
6　Inspection Rules ··· 7
　6.1　Inspection and Acceptance ··· 7
　6.2　Retest and Judgment ··· 7
7　Identification, Transportation and Warranty ·· 7
　7.1　Identification ·· 7
　7.2　Transportation ·· 8
　7.3　Warranty ·· 8

Foreword

This standard is the amended edition of TB/T 2635-1995 *Technical Specification for Heat Treated Rails* by referring to foreign technical standards for heat-treated rails.

The main changes with respect to TB/T 2635-1995 *Technical Specification for Heat Treated Rails* are listed below:

——Terms and definitions are added;

——Code No. for heat treated rails are added;

——New heat treated rail types U75V and U76NbRE are added;

——Chemical composition for rail steel U71Mn is adjusted;

——Hardness scope for rail head section is added and sampling locations for rail head chin are increased;

——Indicators for yield strength and section shrinkage ratio in tensile test are canceled;

——Drop weight test is canceled;

——Twist tolerance for heat treated rail end before straightening is added.

This standard is proposed and managed by Standards and Metrology Research Institute of the Ministry of Railways.

This standard is jointly drafted by Wang Shuqing, Zhou Qingyue, Zhu Mei, Zhan Xinwei, Xu Quan and Hong Haifeng.

Drafting organizations: Metal and Chemical Research Institute, Standard & Metrology Research Institute, Research and Development Center of China Academy of Railway Sciences, Pangang Group Company Ltd. and the rail welding section of Hohhot Railway Administration.

This standard is first issued in March 1995 and this is the first amended edition.

Technical Specification for Heat Treated Rails

1 Scope

This standard stipulates the technical specifications, test method, inspection rules, etc. for heat treated rails (including off-line heat treatment and on-line heat treatment).

This standard is applicable to full length heat treatment of 50 kg/m-75 kg/m rails of U71Mn, U75V, U76NbRE. Rails of other types can take it for reference.

2 Normative References

The following referenced documents are indispensable for the application of this standard. For dated references, only the edition cited applies, while the following modifications (excluding corrigendum) and revisions do not apply. However, all parties reaching an agreement based on this standard are encouraged to study whether the latest edition of these documents shall apply. For undated references, the latest edition of the referenced document (including all amendments) applies.

GB/T 222   *Permissible Tolerances Chemical Composition of Steel Products*
GB/T 223   *Methods for Chemical Analysis of Iron, Steel and Alloy*
GB/T 228-2002   *Metallic Materials-tensile Testing at Ambient Temperature*
GB/T 230-1991   *Testing Method for Metallic Rockwell Hardness*
GB/T 231.1-2002   *Metallic Materials-brinell Hardness Test Part 1: Test Method*
GB/T 13298-1991   *Metal-inspection Method for Microstructure*
TB/T 2344-2003   *Technical Specifications for the Procurement of 43 kg/m-75 kg/m as Rolled Rails*

3 Terms and Definitions

The following terms and definitions apply for the purposes of This standard.

3.1
Off-line Heat Treatment

Rails are hardened to austenitic state by reheating and then cooled down to fine lamellar pearlite.

3.2
On-line Heat Treatment

Rails are directly cooled down to fine pearlite after rolling without being reheated to austenitic state.

4 Technical Specifications

4.1 Rails

Rails for full length heat treatment shall be conformed to regulations of TB/T 2344-2003, with those severely rusted not allowed for heat treatment.

4.2 Chemical Composition

4.2.1 Code No., steel grade and chemical composition (melting analysis) shall be conformed to regulations in Table 1.

4.2.2 Chemical composition for finished rail product shall be limited within the scope stipulated in Table 1, with allowable tolerance conformed to regulations of GB/T 222.

Table 1 Chemical Composition

Steel grade	Chemical composition %							
	C	Si	Mn	P	S	V	Nb	RE[a]
U71Mn(C)[b]	0.70-0.76	0.15-0.35	1.20-1.40	≤0.030	≤0.030	-	-	-
U75V	0.71-0.80	0.50-0.80	0.70-1.05	≤0.030	≤0.030	0.04-0.12	-	-
U76NbRE	0.72-0.80	0.60-0.90	1.00-1.30	≤0.030	≤0.030	-	0.02-0.05	0.02-0.05
[a] added quantity of RE								
[b] adjusted composition for U71Mn in TB/T 2344-2003								

4.3 Profile of Hardened Layer

The hardened layer of rail head after off line heat treatment shall be shaped as a cap symmetrically covering the rail head, as shown in Figure 1.

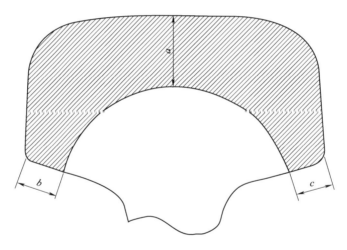

Figure 1 Profile and Depth of the Hardened Layer of Rail Head

4.4 Depth of the Hardened Layer

Depth of the hardened layer of rail head after off line heat treatment refers to the depth heated to austenitizing temperature. The area in black after immersion is shown in Figure 1, in which:

1) $a \geqslant 15$ mm;
2) $b \geqslant 10$ mm, $c \geqslant 10$ mm.

4.5 Hardness

4.5.1 Hardness of rail head section and its distribution

Rockwell hardness of rail head section is measured for locations shown in Figure. 2 and the results shall be conformed to regulations in Table 2. The hardness values shall equably transit from the rail surface to inside and drastic change of hardness is not allowed.

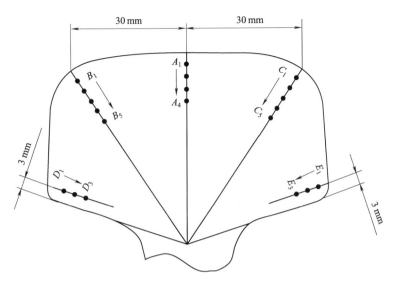

The first sampling point for hardness measurement is 3 mm from the rail surface and then clearance between every two sampling points are all 3 mm.

Figure 2 Locations for Hardness Measurement

Table 2 Hardness of the Hardened Layer at Rail Head Section

Code No.	Steel grade	Rockwell hardness HRC			
		A_1, B_1, C_1	A_4, B_5, C_5	D_1, E_1	D_3, E_3
H340	U71Mn(C)	36.0-42.0	≥32.5	≥35.0	≥32.5
	U75V				
	U76NbRE				
H370	U75V	37.0-43.0	≥34.0		
	U76NbRE				

4.5.2 Hardness of rail head top surface and its fluctuation

Brinell hardness is measured at top surface center of rail head and the result shall be conformed to regulations in Table 3. At the same location, Brinell hardness is measured again after removing the decarburization (0.5 mm). Hardness of the whole rail should vary within 30 HBW.

Table 3 Hardness of Rail Head Top Surface

Code No.	Steel grade	Brinell hardness HBW10/3000
H340	U71Mn(C)	332-391
	U75V	
	U76NbRE	
H370	U75V	341-401
	U76NbRE	

4.6 Microstructure

Microstructure of the hardened layer should be fine lamellar pearlite. No martensite or bainite is allowed to be found in heat-treated rail and little ferrite dispersedly distributed is allowed.

4.7 Tensile Performance

Tensile performance of the hardened layer shall be conformed to regulations in Table 4.

Table 4 Tensile Performance

Code No.	Steel grade	Tensile strength R_m N/mm²	Elongation rate A %
H340	U71Mn(C)	≥1 180	10
H340	U75V	≥1 180	10
H340	U76NbRE	≥1 180	10
H370	U75V	≥1 230	10
H370	U76NbRE	≥1 230	10

4.8 External Shape and Allowable Tolerances

4.8.1 The heat treated rails shall be straight after straightening without wave bending, manual bending or apparent twist. Rails with unqualified straightness are allowed for further straightening, while straightening roll could only be used for once.

4.8.2 External shape, dimension and allowable tolerances of dimension for heat treated rails shall be conformed to regulations of TB/T 2344-2003, while the allowable bending and twist value shall be conformed to regulations in Table 5.

Table 5 Allowable Bending and Twist

Items		Allowable tolerance	Requirement
Bending at rail end	up and down	≤250 mm	before straightening
Bending at rail end (within 1 m from rail end)	up	≤0.5 mm/1 m	after straightening
Bending at rail end (within 1 m from rail end)	down	≤0.2 mm/1 m	after straightening
Bending at rail end (within 1 m from rail end)	left and right	≤0.5 mm/1 m	after straightening
Bending for full length (excluding the part 1 m from rail end)	vertical	≤0.4 mm/1 m	after straightening
Bending for full length (excluding the part 1 m from rail end)	horizontal	≤0.7 mm/1.5 m	after straightening
Twist for the whole rail		≤1/10 000	after straightening

4.9 Surface Quality

4.9.1 No crack, over-burning or local melting shall be found at any location on surface of heat treated rails.

4.9.2 Depth of longitudinal or lateral nick or crack of heat treated rails shall be conformed to regulations in TB/T 2344-2003.

5 Test Method

5.1 Sampling

5.1.1 Heat treated rails for 1 m, either cut from rails with the same chemical composition and then heat treated with the same technique, or directly cut from heat treated rails excluding the part 0.5 m from rail end, are selected as test samples.

5.1.2 Samples for tests of hardened layer profile, depth, hardness of rail head section, surface hardness of rail head, tensile performance and microstructure shall be cut from central testing rails or heat treated rails excluding the part 0.5 m from rail end.

5.2 Testing Content, Sampling Locations and Quantity of Heat Treated Rail Samples

Major testing content, sampling locations and quantity of heat treated rail samples shall be conformed to regulations in Table 6 and other testing content shall be conformed to regulations of TB/T 2344-2003.

Table 6 Testing Content, Sampling Locations and Sampling Quantity

No.	Items	Sampling location	Quantity
1	Chemical composition (melting analysis)	GB/T 222	1 per furnace
2	Profile of hardened layer	Rail head section	1
3	Depth of hardened layer	Rail head section, using the sample for item 2	1
4	Surface hardness of rail head and variation	See 5.2.2	1 1 piece/12 months
5	Hardness of rail head section	Rail head section, using the item 2 sample	1
6	Microstructure	Rail head section, using the item 2 sample	1
7	Tensile performance	Rail head, see Figure 3	2
8	External shape	Full length	Each rail
9	Surface quality	Full length	Each rail

5.3 Test for Chemical Composition Analysis

While making the chemical composition analysis for finished heat-treated rails, the sampling locations shall be conformed to regulations in section 6.2 of TB/T 2344-2003. Test could be done according to regulations in GB/T 223.

5.4 Test for Profile and Depth of the Hardened Layer

5.4.1 One piece of rail shall be taken for testing for every 2 km (track km, same below) or less heat treated rails.

5.4.2 Rail head 15 mm-20 mm thick is cut as samples from central testing rail or heat treated rail excluding the part 0.5 m from rail end.

5.4.3 Samples for rail head section are dipped in 10% nitric acid alcohol solution after abraded and polished with metallographic abrasive paper. An symmetrically distributed hardened layer profile in black are expected to be presented and depth of the hardened layer shall be measured.

5.4.4 Depth and profile of the hardened layer of rails with on-line heat treatment will not be tested.

5.5 Hardness Test

5.5.1 Test for hardness of rail head section and its distribution

5.5.1.1 One piece of rail shall be taken for testing for every 2 km or less heat treated rails.

5.5.1.2 Rail head 15 mm-20 mm thick is cut as samples from central testing rail or heat treated rails excluding the part 0.5 m from rail end.

5.5.1.3 Hardness test is done with samples for rail head section at locations shown in Figure 2, with testing method conformed to regulations of GB/T 230-1991.

5.5.2 Hardness of rail head top surface and its fluctuation

5.5.2.1 One piece of rail shall be taken for testing for every 2 km or less heat treated rails.

5.5.2.2 Rail sample no less than 150 mm shall be cut from central testing rail or heat treated rail excluding the part 0.5 m from rail end to test hardness of rail head top surface and its fluctuation.

5.5.2.3 Rail samples of 150 mm shall be cut respectively at 0.5 m from the rail end and every 5 m of a 25 m long heat treated rail for hardness fluctuation test of rail head top surface every 12 months.

5.5.2.4 After removing 1 mm from rail head top surface, Brinell hardness is tested every 20 mm along the longitudinal center line, with the testing method conformed to regulations of GB/T 231.1-2002.

5.6 Tensile Test

5.6.1 One piece of rail is taken for test for every 10 km off-line heat treated rail or every 1 furnace on-line heat-treated rails.

5.6.2 Rail head of 100 mm is cut from central testing rail or heat treated rail excluding the part 0.5 m from rail end for tensile test. The sampling locations are shown in Figure 3. Diameter of rail head sample $d_0 = 6$ mm and the length $l_0 = 5\,d_0$.

5.6.3 Method for tensile test shall be conformed to regulations of GB/T 228-2002.

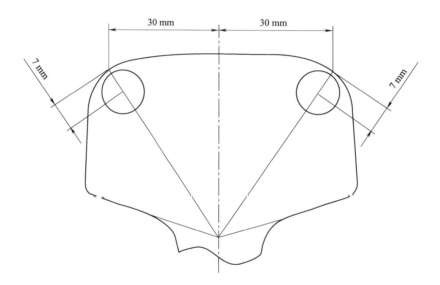

Figure 3　Sampling Locations

5.7 Microstructure Test

5.7.1 One piece of rail shall be taken for testing for every 2 km or less heat treated rails.

5.7.2 Rail head 15 mm-20 mm thick is cut as samples from central testing rail or heat treated rail excluding the part 0.5 m from rail end.

5.7.3 Test for microstructure is done on the rail head section sample, with the test method conformed to regulations in GB/T 13298-1991.

5.8 Inspection for External Shape and Surface Quality

5.8.1 Inspection for external shape and surface quality shall be done one by one for each batch of heat treated rails.

5.8.2 Leveling ruler and feeler gauge especially for rail inspection shall be used to evaluate rail bending at both vertical and horizontal directions.

5.8.3 Surface quality shall be visually inspected.

6 Inspection Rules

6.1 Inspection and Acceptance

6.1.1 Heat treated rails shall be inspected by batch. Each batch is composed of certain number of rails with the same rail type, steel grade and heat treatment technique.

6.1.2 Inspection and acceptance of heat treated rails shall be done by quality inspection department of the manufacturer. The purchaser is entitled for random inspection according to regulations of this standard when necessary.

6.1.3 Inspection of heat treated rails include chemical composition, profile and depth of hardened layer, hardness of rail head section, hardness of rail head top surface and variation, tensile performance, microstructure, external shape and surface quality. The inspection result shall be conformed to regulations in 4.2-4.9.

6.1.4 Heat treated rails qualified after inspection shall be attached with quality certificate, indicating manufacturer, rail type, steel grade, heat treatment plant, code No., manufacturing date, inspector as well as the inspection data.

6.2 Retest and Judgment

6.2.1 Profile and depth of the hardened layer, hardness of rail head section and tensile performance

When the initial inspection result is unqualified, two rail samples will be taken from two heat treated rails randomly selected in the same batch for retest. In case that the test results for the two samples are conformed to regulations of this standard, the batch of heat treated rails are considered as qualified.

If one of the two samples is found inconsistent with this standard, another two heat treated rails will be randomly selected in the same batch for inspection. The batch of heat treated rails will be considered as qualified if the two samples proved to be consistent with this standard and unqualified if one of the two samples found inconsistent with this standard.

6.2.2 Microstructure

When the microstructure is found unqualified, re-inspection is not allowed and the batch of heat treated rails is considered as unqualified.

6.2.3 When the profile and depth of the hardened layer, hardness of rail head section, tensile performance and microstructure are found unqualified, further heat treatment is allowed for the batch of rails. In case that no harmful damage like martensite, bainite, etc. is found and the rails are not heat treated, the batch of rails could be delivered as non-heat-treated rails and labeled correspondingly.

7 Identification, Transportation and Warranty

7.1 Identification

The following contents shall be marked at rail web 1 m from rail end:
a) heat treatment plant;
b) code No. for heat treated rails;
c) month and last two figures of year for heat treatment;
d) serial number.

The heat treated rails shall be identified as follows:

7.2 Transportation

Shipping of heat treated rails shall be conformed to relevant regulations concerning rail transportation.

7.3 Warranty

Warranty of heat treated rails shall be conformed to relevant regulations in TB/T 2344-2003.